ALLEN PARK PUBLIC LIBRARY #4
8100 Allen Road
Allen Park, MI 48101-1708
313-381-2425

TWEEN 591.563S

ANIMAL BEHAVIOR REVEALED

HOW ANIMALS PLAY

REBECCA STEFOFF

Cavendish Square
New York

This book is dedicated to LINUS BERKELEY GEORGE BLACKHEART

With special thanks to Dr. Michael D. Breed of the Environmental, Population, and Organismic Biology department at the University of Colorado, Boulder, for reviewing the text of this book.

Published in 2014 by Cavendish Square Publishing, LLC
303 Park Avenue South, Suite 1247, New York, NY 10010

Copyright © 2014 by Cavendish Square Publishing, LLC

First Edition

No part of this publication may be reproduced, stored in a retrieval system, or transmitted in any form or by any means—electronic, mechanical, photocopying, recording, or otherwise—without the prior permission of the copyright owner. Request for permission should be addressed to Permissions, Cavendish Square Publishing, 303 Park Avenue South, Suite 1247, New York, NY 10010. Tel (877) 980-4450; fax (877) 980-4454.

Website: cavendishsq.com

This publication represents the opinions and views of the author based on his or her personal experience, knowledge, and research. The information in this book serves as a general guide only. The author and publisher have used their best efforts in preparing this book and disclaim liability rising directly or indirectly from the use and application of this book.

CPSIA Compliance Information: Batch #WS13CSQ

All websites were available and accurate when this book was sent to press

Library of Congress Cataloging-in-Publication Data
Stefoff, Rebecca.
How animals play / Rebecca Stefoff. • p. cm.—(Animal behavior revealed)
Includes bibliographical references and index.
Summary: "Provides comprehensive information on the science behind how and why animals play"—Provided by publisher.
ISBN 978-1-60870-512-2 (hardcover) • ISBN 978-1-62712-023-4 (paperback) • ISBN 978-1-60870-614-3 (ebook)
1. Play behavior in animals—Juvenile literature. I. Title. II. Series.
QL763.5.S74 2013 • 591.56'3 • 2010053241

Art Director: Anahid Hamparian • Series Designer: Alicia Mikles

Photo research by Laurie Platt Winfrey, Carousel Research, Inc.

The photographs in this book are used by permission and through the courtesy of:
Cover: Minden Pictures/ Ernie Janes; *Alamy*: Robert Shantz, 15; Robert Harding Picture Library, 21; Goncalo Diniz, 29; All Canada Photos, 32; Juniors Bildarchiv, 65; *Associated Press*, 7; *Gordon M. Burghardt*, 40; *Getty Images*: Louis Mazzatenta/National Geographic, 23; Peter Macdiarmid, 53; *Minden Pictures*: Ahup Shah, 19; Hiroya Minakuchi, 31; Andrew Walmsley, 51; Kim Taylor, 55; Seiichi Meguro, 62; Katherine Feng, 69; Stephen Kazlowski, 72; *National Geographic Images*: Norbert Rosing, 4, 35; Kim Wolhuter, 26; Ingo Arndt/Minden Pictures, 37; *National Zoological Park/ Smithsonian Institution:* 42; *Newscom:* Dallas + John Heaton Stock, Connection USA, 49; Jorge Guerrero/ Getty Images, 60; *Photo Researchers:* Toni Angemayer, 39; Carolyn McKeone, 47; *Superstock:* Science Faction, 10; Marka, 13; NHPA, 25; National Geographic, 57.

Printed in the United States of America

CONTENTS

Investigating Animal Behavior	**5**
1. Dinosaurs at Play	**11**
2. What Is Play?	**27**
3. Animal Games	**43**
4. The Meaning of Play	**63**
Glossary	**74**
Find Out More	**76**
Bibliography	**77**
Index	**79**

INVESTIGATING ANIMAL BEHAVIOR

Wolf pups spend a lot of time and energy running around, pouncing on anything that moves, and pretending to fight with each other. Like tiger cubs, dolphins, parrots, and many other animals, young wolves are playful. Most of us will never romp with a wolf, but every day millions of people play with wolves' close relatives, dogs. People also like to watch animals play—in nature documentaries and amateur videos, in zoos, or in yards and parks. Play is one way that we humans connect with the creatures that share our world.

People have always been fascinated by animals. Tens of thousands of years ago, our ancestors painted lifelike pictures of bears, bison, and deer on cave walls. Twenty-five centuries ago, Greek thinkers wrote about animals and their habits. Those writings were the beginning of **zoology**, the scientific study of animals.

Wolf pups in Montana greet each other. Young wolves and other predators play by nipping, wrestling, and pouncing on each other—gentle versions of the moves that adults use for fighting and hunting.

HOW ANIMALS PLAY

In time zoologists wondered why animals do the things they do. Do wolf pups play because of **instinct**—a built-in pattern of behavior that is programmed into the wolves' genes? Or do they learn to play from their mother, their older siblings, and other adult wolves in their pack? And *why* do they play? Has the baby learned to be afraid because its mother and the other monkeys in its troop scream and run when they see a snake? Questions like these led to **ethology**, the branch of zoology that studies animal behavior.

Ethology became established as a science in the twentieth century. One of its pioneers was Konrad Lorenz of Austria, who studied the behavior of geese and ducks. When these birds hatch, they usually see their mother right away, and after that they follow her around. Lorenz wondered what would happen if young birds hatched apart from others of their **species**. He experimented with geese and found that newly hatched birds bonded with the first thing they saw. Young geese that bonded with Lorenz followed him around as if he were their mother. Lorenz called this behavior imprinting.

Lorenz published his findings in 1935. The next year he met Nikolaas Tinbergen, a Dutch zoologist who was also curious about how animals react to signals from their environment. Together Tinbergen and Lorenz studied seagulls and ducks. They discovered that young birds raised by humans showed no fear of round or square cardboard cutouts, but they instinctively recognized dangerous shapes. When cutouts shaped like hawks and eagles flew over their nests, the birds became afraid, even though they had never seen those **predators**.

INVESTIGATING ANIMAL BEHAVIOR

Konrad Lorenz spent much of his life surrounded by geese and ducks that viewed him as their "mother." Lorenz began studying these birds as a young man and later won a Nobel Prize for his discoveries about their behavior.

HOW ANIMALS PLAY

The work of Lorenz and Tinbergen was a step toward understanding animal instincts. In 1973 the two men and Austrian bee expert Karl von Frisch shared a Nobel Prize, one of the highest scientific honors, for their work in the new science of animal behavior.

Tinbergen came up with four key questions to ask about the things that animals do. Today ethologists and animal behavior researchers still use those questions to guide their investigations. The questions are:

- What causes the animal's behavior? A scientist who sees a fox pounce on a pile of dead leaves, for example, would keep watching in order to see if the fox is hunting for food, such as a mouse, or is playing.
- Does the behavior change over the animal's lifetime? Do young animals play more than old ones do, or in different ways?
- How does the animal's behavior compare with the way similar species behave? Do dogs, wolves, and coyotes play in the same ways?
- Does the behavior help or hurt the animal's chances of surviving and reproducing? If kittens spend a lot of time stalking and jumping on each other, will they be better hunters when they grow up?

Tinbergen also warned other scientists against the tendency toward **anthropomorphism**, which is a fancy way of saying "giving human qualities to animals." When we describe animals in human terms, such as saying, "Oh, that bear is sad," we are anthropomorphizing. For a long time, anthropomorphism was strictly forbidden in ethology. In recent years, though, scientists have learned much about the inner lives of animals—how they think, feel, communicate, and play. Thinking that animals are completely different from people may be as big a mistake as thinking that animals are just like people.

The study of animal behavior takes many forms. Some researchers focus on **psychology**, the study of human and animal minds, or on **evolution**, the history of life on Earth. **Sociobiologists** study animals that live in social groups, such as ants and prairie dogs. Behavioral ecologists look at how animals interact with their environments. Other researchers investigate animal communication and intelligence.

This book looks at animal play—a subject that goes far beyond kittens chasing pieces of string. How do we know when an animal is playing? Which animals play, and which do not? What is the purpose of play? To answer these questions, scientists are making new discoveries about animal behavior. They are helping us understand what play means in animals' lives, and in our own.

1. DINOSAURS AT PLAY

One spring morning 75 million years ago, the sun rose over the vast inland sea that covered most of what is now Texas. The western shore of that ancient sea is now a land of deserts and dry plains. Back then, though, it was a patchwork of forests, grasslands, and wetlands, and it teemed with life—especially dinosaurs.

Bigheaded, clumsy *Tyrannosaurus rex* babies wrestled each other in their nest like puppies. Three-horned triceratopses leaped and frisked in the sunshine. Swift young velociraptors spun themselves dizzy trying to catch their own tails, and an armor-plated ankylosaurus carried its favorite plaything—a small, chewed log—in its mouth as it trudged across a meadow.

Did such things really happen? Did the dinosaurs play?

That question is part of a larger question: When and why did play appear in the animal world? Before animal behavior researchers can investigate those questions, they must decide what play is.

Fearsome hunting hooks gave *Deinonychus* dinosaurs their name, which means "terrible claw." These predators lived in North America 100 million years ago. Did their young play together as wolf pups do?

You might think that nothing could be simpler or easier to understand than playing, but scientists have not yet come up with a definition that everyone accepts. Scientists do not yet know why animals (and people) play, either. Most researchers think that play helps animals in some way, such as giving them a chance to practice skills or develop muscle strength. The experts agree, though, that we have a lot more to learn about how animals play.

The Science of Play

The first studies of animal play were simple descriptions of animals having what looked like fun. By the twentieth century, scientists were asking questions about play and trying to answer them. Researchers today look at play from a scientific angle, but studying play is surprisingly difficult.

Early Research: Leaping Lambs and Crouching Cats

In 1851 an Englishman named Edward P. Thompson wrote an early description of animal play. In a book called *The Passions of Animals* he listed examples of playfulness.

"Dogs, particularly young ones," wrote Thompson, get carried away with the spirit of playfulness, "rolling over and chasing each other in circles, seizing and shaking objects as if in anger, and enticing even their masters to join in their games." Lambs gather in groups, "racing and sporting with each other in the most interesting manner." Deer "engage in a sham [pretend] battle, or a trial of strength, by twisting their horns together, and pushing for the mastery."

DINOSAURS AT PLAY

Because kittens and cats instinctively try to catch small objects thrown through the air, owners of these felines can bond with their pets through play.

Thompson pointed out that for some animals, playing resembles hunting for **prey.** He singled out "young cats," which "leap after every small and moving object . . . crouch, and steal forward ready for the spring, the body quivering and the tail vibrating with emotion." He mentioned crabs that "play with little round stones, and empty shells, as cats do with a cork or small ball."

Another nineteenth-century Englishman who made careful observations of animal life was Charles Darwin, whose 1859 book *On the Origin of Species* launched the study of evolution, which is the way new species form over time because of changes in earlier species. Darwin also wrote a book about emotions in animals and people. In it, he connected happiness with play.

Happiness, Darwin wrote, "is never better exhibited than by young animals, such as puppies, kittens, lambs, &c., when playing together, like our own children." Even insects play, Darwin claimed. He hadn't seen bugs playing, but he had read about ants that chased and pretended to bite each other, like puppies, in the work of a Swiss scientist named Pierre Huber.

Studying Play Is Hard Work

In the twentieth century, biologists shifted from describing animal play to trying to explain it. They wanted to know how play came to exist, and what purposes it served in animals' lives. Finding answers has not been easy. Scientists who study animal play face many challenges.

One challenge is the simple difficulty of seeing animals when they are playing. Marc Bekoff, a biologist and ethologist who is a leading animal play researcher, uses the example of two projects to study pronghorn antelope. One project is designed to study how the pronghorn **forage,** or look for food and eat it. The other project is to study their play. The foraging researchers can show up "at just about any time of day," Bekoff explains. Even missing a whole week of observations is no problem because "the animals are always foraging."

Pronghorn antelope live in the American West. One challenge of studying play in this species is that antelope play only when young, during less than a fifth of their lifetimes.

But to study play in pronghorns, Bekoff points out, researchers must be in position by five o'clock in the morning. And because pronghorns play for only a short period early in their lives, a researcher who misses a week has missed about 15 percent of a particular herd's total playtime.

Researchers go to great lengths to observe wild animals in their natural **habitats** when they can, but it is not always easy to find and track lemurs in the forests of Madagascar, or lemmings in the Arctic tundra. Much of the information that scientists have collected about animal play comes from observations of tame or captive animals.

Defining Play

Play fascinates scientists for many reasons. Evolutionary biologists want to know when play first appeared in the history of animal life. Comparative psychologists study the ways animal and human play are different or alike. Sociobiologists are curious about how animals' relationships with other animals shapes play.

No matter what approach scientists take, though, they must try to answer one basic question: What *is* play?

If It's Not Useful Then It Must Be Play

Scientists have had a hard time defining play in either humans or animals. One psychologist wrote in 2000 that "it is unlikely that researchers will ever come up with a satisfactory definition of play." It's true there is no perfect definition, but several experts have tackled the challenge.

THE TROUBLE WITH A GOOD STORY

Suppose a backpacker returned from a camping trip and said that he'd seen five wild grizzly bears playing catch with a pumpkin. His story would be an anecdote, which is simply someone's account of something that happened.

Anecdotal evidence creates problems for scientists. Anecdotes are not always true, or accurately reported. The backpacker could have made up his story, or only imagined he saw the bears. But what if he had pictures or video of them tossing a pumpkin? Scientists couldn't be positive that the images weren't a trick involving trained bears, or really good bear costumes. Now suppose the backpacker is a respectable scientist who knows a lot about bears and has no reason to lie. The bears' game of catch-the-pumpkin would still be a single strange event, not part of an organized scientific research project.

If bears were never again seen playing catch, the backpacker's anecdote would not add much to our scientific knowledge. It might be repeated in books, the way Darwin repeated the story about ants playing like puppies, but it would remain "just an anecdote."

Animal play research involves a lot of anecdotes because many examples of animal play are anecdotal—things seen just once, or only a few times. Anecdotes on their own don't make strong evidence, but they are important. They give researchers ideas for new theories, wildlife research projects, or experiments.

How Animals Play

One approach is to set play apart from the other things animals do. In this kind of definition, play is any physical activity that does not have a clear purpose. In other words, if an animal is doing something, and it isn't eating, hunting for food, defending itself, escaping from a predator, mating, marking its territory, or doing anything else that is essential to surviving and reproducing—it must be playing. The animal's actions when playing, however, may look a lot like its actions when hunting or fighting.

To use one of Thompson's examples from 1851, a young cat that pounces on a windblown leaf looks just like a young cat that pounces on a scurrying mouse, but one is playing and one is hunting. Play is activity that is not related to some other purpose in the animal's life at that moment. Even if playing with a leaf helps the cat become a better hunter, someone watching the cat would not see that long-term benefit because it is separate in time from the playing behavior.

Five Features of Play

Another way to define play is to look at examples of play in humans and animals, then list the features that these behaviors share. This is the method used by Gordon Burghardt, a psychologist, evolutionary biologist, and animal behavior researcher. Play, Burghardt suggests, has these five features:

1. Play is not necessary for immediate survival. Play behavior is not necessary or essential to the animal's survival. Whatever long-term benefits an animal may gain from playing, play does not help the animal survive at the moment it is playing.

DINOSAURS AT PLAY

This lion cub is still nursing from its mother and does not yet eat meat. Pawing and gnawing at prey killed by older lions is a form of playing that is also a rehearsal for life as a hunter.

Lion cubs that tussle with the bones of a dead gazelle, for example, may be trying out moves that they will use as adults when they are hunting, but as cubs they are still getting fed by their mother. They don't yet eat prey, so the bones have no immediate survival value to them.

2. The animal starts the game. Play originates with the animal itself, and it is voluntary. This means that play happens without any outside force acting on the animal, as when horses start running for no particular reason.

From the human point of view, play appears to be interesting, enjoyable, or both, to the animal. But do animals really have fun? In human brains, feelings of pleasure and enjoyment are triggered by a chemical called dopamine. That same chemical is created in animal brains, which makes scientists think that many animals—perhaps most of them—can feel pleasure.

Play is also intentional, or done on purpose. Polar bear cubs might slide down an icy slope once by accident, but if they climb to the top and slide down again, several times, they are playing.

3. Play differs from serious behaviors. An animal's playful actions may look like the actions that the same animal uses in "serious" behavior such as fighting or hunting, but the play version of the behavior is different from the serious version. It is gentler, or more awkward, or performed at a different speed.

When a young kangaroo, called a joey, boxes and spars with its mother, reaching up from her pouch to swat at her face, the mother "fights" back, but only in play. Her movements are slow and gentle enough not to harm her joey. When animal play includes mock or imitation versions of mating, hunting, or fighting, the play ends without actual mating or injury.

If a mother kangaroo boxes or fights with her joey, she will not hurt the youngster. Play-fighting is a form of communication and bonding between mother and offspring.

4. Play is repeated. Play is behavior that an animal repeats more than once—not in exactly the same way each time, but in similar ways. Anyone who has thrown a stick or tennis ball for a dog to fetch knows that a dog may be happy to repeat this play behavior, chasing the thrown object in any direction, until its human's throwing arm wears out.

The repeated nature of play rules out one-time anecdotes such as the imaginary pumpkin-pitching grizzlies. It also rules out exploratory behavior, which can look like play. Exploratory behavior happens when an animal encounters a new environment or object. Many animals are curious about new things. At some point they smell, taste, push, or handle objects, or they run or climb around new settings. If you put a parrot in a new cage that is full of toys, the bird will climb around the cage and move all the toys with its feet and its beak to see what they do or how to use them. This is exploration. If the parrot continues to swing on its moving perch and shake its rattle when it has gotten used to its cage, it is playing.

5. The animal is relaxed. Animals play when they are comfortable, well fed, and free of worry. To be relaxed enough to play, they cannot be trying to meet a basic need such as finding food, shelter, or a mate. They also cannot be trying to avoid being eaten. Sick, hungry, cold, or frightened animals do occasionally play, but such behavior is not common. When it does happen, the play usually lacks energy. Animals under stress typically use all of their energy and attention just trying to stay alive.

Researchers have used the term "relaxed field" to describe a state of physical and mental comfort. When an animal is in a relaxed field, its behavior is easier to recognize as play.

Back to the Dinosaurs

The last of the dinosaurs became extinct about 65 million years ago. Still, we know a lot about how these ancient creatures looked because we have unearthed many remains of their bones, teeth, and sometimes even skin and feathers. These remains became buried, and over long periods of time, minerals from water in the soil turned them to stone. They exist today as fossils, giving us evidence of dinosaur **biology.** But evidence of how the dinosaurs acted is not so easy to find.

Fossils do reveal traces of dinosaur behavior. By studying fossilized footprints, scientists have gained new understanding of how various kinds of dinosaurs walked and ran. Fossilized nests have also given us clues about dinosaur parenting.

Confuciusornis, the oldest known bird to have a beak instead of teeth, lived about 125 million years ago in China, alongside many dinosaur species. Fossils tell us what these long-extinct animals looked like, right down to the pattern of their feathers, but not whether they played.

HOW ANIMALS PLAY

In 2003, for example, scientists working in China discovered a 100-million-year-old nest. The skeletons of thirty-four young dinosaurs and one adult were in or near the nest. Finds such as this one strongly suggest that at least some dinosaurs gave parental care to their offspring, and that parent-child groups stayed together for a while after the young dinosaurs hatched.

No fossils, however, have told us whether or not the dinosaurs played. Scientists must look for clues about the origins of play in the behavior of living animals.

Dinosaurs descended from a group of ancient animals called **archosaurs**. So did a line of large reptiles called **crocodilians.** The dinosaurs died out, but most experts now agree that birds evolved from dinosaurs. In terms of evolution, this means that the dinosaurs' closest living relatives are two groups of archosaurs that still exist today: the birds and the crocodilians, a category made up of crocodiles, alligators, gharials, and caimans.

Birds play, but scientists do not yet know whether *all* kinds of birds play. Crocodilian play is not fully understood, either, although some reptiles play. Because these descendants and relatives of dinosaurs play, it is possible that some of the dinosaurs played, too. How early in the history of life animals started to play remains a puzzle, however. Play may have evolved separately in two or more kinds of animals, the same way that flight evolved separately in insects, birds, and bats. Play may have appeared more than once in the history of life.

Scientists hope to learn more about the origin of play by learning more about play as it exists today. Once they have examined all the

DINOSAURS AT PLAY

Birds play tug-of-war with a feather. The fact that these living relatives of dinosaurs play might mean that dinosaurs played, too, but scientists may never know for sure.

known animal groups for evidence of play, they will have more complete knowledge about which animals play and which do not. By applying this knowledge to what we have learned about evolution—about how various groups of animals split off from earlier groups and developed into new forms—scientists may come closer to pinpointing the beginning of play in the early branching of life's family tree.

2. WHAT IS PLAY?

Two spotted hyenas lope along an African riverbank, taking turns chasing each other, growling, and playing tug-of-war with a stick grasped in their jaws. How would a scientist describe this behavior? Probably by using the same basic classification of play that psychologists use when they are talking about human children.

Classification is one of science's most useful tools. To classify things is to put them in groups based on ways they are alike and different. Fish, for example, can be divided into groups based on the kind of water they live in. A fish is classified as freshwater, saltwater, or able to live in either freshwater or salt water.

Play can also be classified. Based on studies of people and animals at play, many experts divide play into three main categories: **locomotor play**, **object play**, and **social play**. This classification is a vital tool for researchers who observe wild or captive animals and record what they do. By making it easier to identify specific types of play, the classification helps researchers collect information about who plays, and why.

For scientists studying animal behavior, it is not enough to say that these South African spotted hyenas are playing. The scientists use a classification system to record exactly what kind of play the hyenas do, and for how long.

Locomotor Play

Play that involves simply moving the body is called locomotor play. Walking, running, jumping, leaping, head-shaking, kicking, pawing the air, pouncing, and bouncing are examples. This kind of physical play often includes another kind of movement as well—a rotation or twisting of the neck or body. This type of movement occurs when a dog rolls sideways as it leaps into the air, when a bull bucks, or when a horse cranes its neck to look over its shoulder while it prances. The term **locomotor-rotational play**, or LR, describes these actions that involve twisting the body as well as moving it through space.

Locomotor play is especially common in the large group of animals called **ungulates**, which are mammals with hooves. The group includes horses, donkeys, camels, deer, antelope, cattle, goats, sheep, and giraffes. Ungulates often run, jump, climb, and whirl when playing. We sometimes use the word "capering" to describe this kind of lively physical activity in people as well as animals. The term comes from the Latin word *caper*, which means "male goat"—one of the most playful ungulates.

Object Play

Play that involves physical items is object play. The animal may bite, push, carry, hit, or chase an object. Object play is well known to pet owners who make or buy toys for their dogs, cats, parrots, ferrets, turtles, and other animal companions. It is fairly common among livestock and wild animals as well.

Locomotor play involves moving the body—running, jumping, prancing, or kicking, as this giraffe is doing.

Grazing animals, including horses and cows, sometimes pick up sticks in their mouths and either carry the sticks around or throw them. Many species of birds play with twigs, bits of grass, or human-made objects such as bottle caps and keys. Humans and other **primates**, especially apes and monkeys, have fingers and hands that are very well adapted for grasping and manipulating things, including play objects. At least two species of ocean-dwelling mammals make objects out of air in order to play with them. Beluga whales blow bubbles that they bite or knock apart. Bottlenose dolphins blow strings of bubbles into rings, then swim through these bubble rings or manipulate them with their mouths, joining the rings together, breaking them apart, or swirling them around.

A special type of object play is **predatory play**. Many predators play with objects, stalking, biting, and shaking the play objects just as they would live prey. Some predators also appear to play with live prey—catching it and releasing it, only to catch it again. Researchers have difficulty drawing the line between "playing" and "hunting" in behavior such as this.

A wildlife researcher's anecdote shows the similarities and differences between hunting and playing. The researcher watched a six-month-old fox catch and eat a mouse. Soon the fox caught a second mouse. This time the fox played with the mouse's body for a few minutes and then hid it, which is what predators do when they are going to eat their prey later. The fox then caught a shrew—a small insect-eating mammal that is about the size of a mouse, but less desirable as prey because it has a bad smell.

A beluga whale in a Japanese aquarium makes its own toy: a ring of bubbles. Small whales and dolphins have been seen playing with bubble rings both in captivity and in the wild.

HOW ANIMALS PLAY

Predatory play is a special kind of object play in which a predator, such as a fox, toys with an object as though it were a prey animal. This fox is toying with a dead deer mouse.

WHAT IS PLAY?

The researcher thought that the fox was no longer hungry because it had not eaten its second mouse. It did not seem interested in eating the shrew, which it had caught but not killed. Instead, the fox carried the shrew—still alive—to a stretch of open roadway, a place where the shrew could not disappear immediately into the underbrush if it escaped. The fox then jumped and danced around with the shrew for several minutes but did not hurt it. Afterward, the fox carried the shrew back to the place where it had been caught. As soon as the fox opened its mouth and released the shrew, the little animal vanished into a burrow.

The fox's capture of the first mouse was hunting and feeding behavior. The capture of the second mouse looked like hunting with a little play involved. But what about the shrew? Was it prey or a play object? The fox's behavior toward the shrew started out like hunting, then turned into playing. (The shrew probably did not regard it as playing, though. To the shrew, the fox's behavior might have felt more like terrorizing, or bullying.) The same thing happens with servals. These little African wildcats catch mice and rats without hurting them, then push the captured animals into holes or tree stumps so that the servals can "fish" them out with their front paws. Servals play the same fishing game with nonliving objects, such as bits of bark.

Social Play

Play that involves two or more animals at the same time is social play. Common kinds of social play are chasing, wrestling or rough-and-tumble play (such as rolling and jumping on each other), and mock fighting. Social play can also involve behavior that is related to courtship or mating. Young male deer or horses, for example, may climb onto each other in the same way they would approach a female to mate with her. This doesn't mean that two males are trying to mate. Their play simply includes imitating or rehearsing the sexual behavior that is natural for their species.

Social play usually involves two animals of the same species, but not always. When humans play with dogs, for example, they are engaging in social play across species lines. Cross-species play is more commonly seen in tame or captive animals than in wild ones, but some examples have been recorded from the wild.

In 1994 a wildlife photographer named Norbert Rosing was working in Canada when he captured images of one of his sled dogs playing—wrestling, frisking, and rolling around—with a 1,200-pound male polar bear. Neither animal showed aggression toward the other, and neither was hurt. The bear returned every night for a week to frolic with the sled dogs.

Another photographer, Hugo Van Lawick, saw an adult African gazelle play with two young bat-eared foxes. These foxes are predators, but far too small to attack a gazelle—their normal prey is insects and mice. The gazelle chased them in circles, and they chased it, in behavior that appeared playful. Dogs and monkeys have been known to play together. So have dogs and young bears, wolves and bears, horses and goats, and cats and rabbits.

> Wildlife photographer Norbert Rosing captured amazing images of his sled dogs playing with a polar bear, one of the world's most ferocious predators. The bear returned several times to rough house and play-wrestle with the dogs, never harming them. Somehow the bear and the dogs communicated that they all wanted to play.

Role Reversal

When two animals play-fight, one usually plays the role of the "attacker" while the other "defends" itself. In species that engage in a lot of social play, such as monkeys and rats, two animals may take turns in each role. One squirrel monkey chases or jumps on another, for example, but later the second monkey chases or jumps on the first.

These shifts from one role to the other are called role reversals. They may happen within a single play session, in different sessions on the same day, or on different days. Role reversal is a form of taking turns, or sharing, in social play.

Self-Handicapping

Animals that play together are not always evenly matched. One may be older, stronger, faster, or bigger than the other. Young monkeys, for example, sometimes play with their mothers, aunts, or older siblings. In these sessions, the younger, smaller monkey could quickly "lose" a chase or a play-fight.

To make up for its size advantage, the larger monkey may crouch, slow its movements, or limit its energy and skill. This not only prevents the larger monkey from hurting the smaller animal, it prevents the smaller monkey from being discouraged or frightened, which means that the play will continue. The larger monkey's action is an example of **self-handicapping**. This happens when an animal of any species deliberately gives itself a handicap, or disadvantage, so that it and another animal can play as equals.

WHAT IS PLAY?

Teasing and Bullying

True social play requires that all of the animals taking part are doing so voluntarily, because they want to. For example, if one tiger cub, or a group of cubs, tries to engage another cub in play against its will, the first cub's behavior is more like teasing than playing. If the second cub still does not join in and play, the first cub either loses interest or becomes more aggressive.

Among primates such as these gelada baboons in Ethiopia, play sometimes turns into bullying.

Bullying, when one or more animals pick on a single individual, occurs in the animal world. Primatologists see many examples of this behavior in apes and monkeys, the animals that are most closely related to humans. Occasionally a bullied animal will later become **dominant**, or high-ranking in the group, and turns into a bully. Usually the bullied animal remains **subordinate**, or low-ranking. Sometimes animal bullies seriously injure or even kill their victims. Studies of wild baboons have shown, however, that when food becomes scarce for these monkeys, the bullies in the troop may become warriors, protecting resources for their own community by driving away other troops.

Parallel Play

Parallel play is a version of social play that is seen in very young human children, as well as animals. In parallel play, two or more animals play at the same time, not far from each other. They appear to be aware of each other. They may even watch each other and imitate each other's behavior, but they don't touch, communicate, or interact directly. This kind of behavior is most often seen in social animals—species that spend their lives in groups, such as zebras and prairie dogs. Parallel play can shift quickly into social play and then back again as animals come together and drift apart while playing.

Mixing It Up

A lot of play behavior blends several classes of play, as in the case of the two hyenas on the riverbank. They were engaging in locomotor play (running), object play (with a stick), and social play (together) all

PUTTING ON THE PLAY FACE

A few animals use a rare kind of body language called metacommunication, or "communication about communication." Any signal that changes the meaning of other signals is an example of metacommunication.

If you say to your friends, "I'm going to tell you a story" or "Listen to this great joke," you are using metacommunication. Your words tell your friends that the next thing you are going to say isn't real or true. It's made up.

Animals use metacommunication, too. They don't tell stories or jokes (as far as we know), but they do communicate about communication through body language, actions, and facial expressions. A wolf that crouches as if bowing, with its forelegs stretched out on the ground, its rump raised, and its tail wagging, is giving an invitation to play. Animal researchers say that the wolf is giving a "play bow." The play bow is a signal that tells the other wolves that the actions that follow are going to be a game.

Biting, head-shaking, chasing, and pouncing are normally aggressive actions among wolves. If the play signal appears first, though, the meaning of these gestures changes and they are seen to be friendly. Lions and coyotes use the play bow to send the same message. Anyone who has lived with a frisky dog has seen the play bow, too.

Chimpanzees and some other primates have a different play signal. To invite other animals to play, or to signal that they want to play, they put on an expression called a "play face" that other animals in their group will recognize. A chimp's play face is relaxed, with the mouth open and the top teeth covered by the upper lip. The chimpanzee may also pant softly. Animal researchers call this sound "laughter"—which fits right in with play.

A chimpanzee puts on the "play face," a sign that it wants to play. This signal tells other chimpanzees that a sudden pounce, for example, is the start of a game, not an attack.

HOW ANIMALS PLAY

A sink and mirror become a playground for a black bear cub raised by animal researcher Gordon Burghardt.

at the same time. Most scientists would call this behavior social play, or possibly social play with an object. They would use the label locomotor play for an animal playing by itself.

Many play sessions include more than one type of play. Researcher Gordon Burghardt, who raised two black bear cubs, reported that the bears liked to play after eating their afternoon meal. A typical play session started with each bear doing individual locomotor play, such as rolling or climbing on things. One bear then started playing with a favorite set of toys—a purse containing small objects. This caught the eye of the other cub, which started competing for the objects. Before long the two bears were play-boxing, wrestling, and chasing each other. After a few minutes they returned to the scattered toys. In parallel play, each bear sat on its own, handling one or more of the objects. The cubs had cycled through all the classes of play in a short time—something that human children also do.

Burghardt's bears were mammals, members of the large category of animals that also includes humans and many more. People have a long history of relating to other mammal species—our livestock, work animals, and most of our pets are mammals. Some mammals even have facial expressions that people learn to "read." Studying play in birds, snakes, and squids can be more of a challenge. To learn more about how play is spread across the animal kingdom, though, some scientists are doing just that.

3. ANIMAL GAMES

Pigface lived for more than fifty years in the National Zoo in Washington, D.C. His nickname was not an insult—it was descriptive. His nose stuck out like a pig's snout. Pigface was a Nile soft-shelled turtle, a species that is native to rivers and lakes throughout much of Africa. In 1986 Pigface's keepers started putting objects in his tank to see what he would do with them. They gave the turtle sticks, basketballs, and a floating hoop made out of garden hose. He played with them—a lot.

To find out just how much Pigface played, scientists videotaped his tank and then made an **ethogram**. Animal behavior researchers often record what they see in the form of ethograms, charts that show how an animal spends its time. The chart is divided into categories such as sleeping, resting, eating, grooming, fighting, and playing. (For a more detailed study of play, a researcher might create separate categories for locomotor play, object play, and social play.) The researcher watches

Pigface, as this turtle was affectionately known to keepers and visitors at the National Zoo, surprised scientists by showing that at least one reptile had a strong sense of playfulness.

the animal, either directly or in recordings, and fills in the chart—how often the animal performs each behavior, and for how long. A series of ethograms reveals the animal's activity budget, or how it spends its time.

Pigface spent more than 29 percent of his active time swimming, almost 8 percent standing, 7 percent walking, and 3 percent floating. He spent nearly 21 percent of his time interacting with the objects in his tank, including the fill hose that the keepers used to add freshwater.

The turtle followed objects as they floated around his tank. He nudged them with his snout. He bit, grasped, chewed, pulled, or shook them with his mouth, sometimes using his front legs to pull the items closer or to hold them down.

Pigface also used flowing water as a toy or source of enjoyment. When his tank was being refilled he positioned himself so that the stream of water flowed over his head. If he did not like the way the water felt, he would move the hose until it was aimed to his satisfaction, and then he would remain still until the water stopped flowing. Nile soft-shelled turtles often live in rivers or streams with currents, but the water in Pigface's tank didn't move, except when flowing into the tank from the fill hose. Perhaps this explains why Pigface seemed to enjoy running water.

He engaged in social play, too—not with other turtles, because he was alone in his tank, but with his keeper. When the keeper lowered the fill hose into the tank Pigface would seize it in his mouth and pull it. If the keeper raised the hose (and the turtle), Pigface would try to swim backward, as if playing tug-of-war.

ANIMAL GAMES

One captive turtle playing with objects provided by its human keepers does not prove that play is a natural behavior for turtles. Anecdotes about Pigface show, though, that turtles *do* play. Scientists are now investigating play in a wide range of species—not just turtles but animals of all sorts.

"Incredibly," wrote biologists Marc Bekoff and John A. Byers in 1998, "we still do not know if most mammals do or do not play." Knowledge of play in other kinds of animals was even weaker. Since that time, researchers have learned much more about animal play, but many questions remain to be answered.

Mammals

Biologists and psychologists used to think that only mammals played. Scientists now know that other animals play, too, but play is best known in mammals. Of the 5,500 or so known species of mammals, some have been studied more than others by play researchers. Most of the world's most playful animals—including monkeys, dolphins, rats, dogs, cats, and hoofed animals—belong to a handful of mammal subgroups. Play in other mammals, such as anteaters, armadillos, and opossums, is less well known.

Big Players: Dogs, Cats, Monkeys, Rats, and More

Some subdivisions of the mammal group spend more time playing, and have more complex forms of play, than others. These subdivisions are the primates; the carnivores (animals that eat meat, such as cats, dogs, bears, and raccoons); the proboscids (elephants); the **cetaceans** (whales, porpoises, and dolphins); the pinnipeds (seals and sea lions, whose natural love of play makes them easy to train for animal entertainments at ocean parks); the rodents; and the ungulates.

Not all species are equally playful. Among the rodents, for example, house mice run and hop and also play with objects, while rats, squirrels, and prairie dogs spend more time in social play, especially chasing and wrestling each other. But all three kinds of play—locomotor, object, and social—appear in each of these mammal subdivisions.

Marsupial Mammals

About three hundred species of mammals belong to a separate classification called the marsupials. Unlike other mammals, marsupials give birth at a very early stage in the development of their young. The newborn marsupials continue developing inside a pouch on the mother's body until they are ready to come into the world. Some marsupial species live in the Americas, but most are native to Australia and nearby islands.

The ancestors of marsupials split off from the ancestors of other mammals at least 100 million years ago. Still, marsupial play seems very similar to the play of other mammals, although it is not yet as well studied.

Wombats are stout, burrowing marsupials that frequently engage in locomotor play, such as leaping into the air and standing on two legs. Bandicoots appear to enjoy leaping, burrowing, and climbing. Some of the most playful marsupials are the highly energetic, carnivorous, black-and-white Tasmanian devils. They are among the few marsupials

THE TUBE-HEADED HEDGEHOG

Hedgehogs are small, insect-eating mammals that are sometimes kept as pets. Even though these animals are fairly common and well-known, there are no scientific reports of hedgehog play. One researcher, however, shared an anecdote that hints at hedgehog playfulness, at least in captivity.

A female hedgehog was free to roam on a kitchen floor. After the cardboard tube from a roll of toilet paper was placed on the floor, the hedgehog wedged her head into one end of it. She walked around with the tube on her head until the tube knocked against a wall and fell off. The tube-on-the-head episode could have been an accident, but for the next fifteen minutes the hedgehog kept shoving her head back into the tube, waving the tube around, knocking it off, and moving it so that she could get into it again. She repeated this behavior later the same day and on other days, showing no sign of hostility or fear toward the tube. It's hard to imagine that she was doing anything other than playing.

Pet hedgehogs vary, according to their owners. Some show no interest in toys, while others occasionally push a ball or crawl around with a toilet-paper tube. Play behavior is not yet well studied in many small wild animals.

that play with objects—these devils have been known to pick up and shake toys. They are also famous for their vigorous play-fighting, which is usually accompanied by ear-splitting shrieks and yowls.

Play-fighting is the most common social play among marsupials. Kangaroos and their smaller relatives, wallabies, play-fight by standing on their hind legs and punching each other with their front legs. This behavior looks so much like the human sport of boxing that, in countries without laws to protect animals, kangaroos have been forced to "compete" with human boxers to entertain audiences. When kangaroos and wallabies fight for real, their most deadly weapon is a powerful kick from one of their large, clawed hind feet. During play-fighting they do not kick. This self-handicapping lets the animals play without risking serious injury.

Play-fighting is the only kind of play that has been recorded for the marsupial opossums of the Americas, and it is known in just one species. The gray, short-tailed opossum is a tree-dwelling marsupial about the size of a mouse, native to South America (and widely used in laboratory studies of marsupials). These animals play-box with their front paws while standing on their hind legs.

Birds

A few types of birds, including ostriches and cranes, do not seem to play at all. Otherwise, play appears to be fairly widespread among birds, but sometimes the evidence for it is slim. The single report of hummingbird play, for example, tells of an Anna's hummingbird

Among kangaroos and their marsupial relatives the wallabies, juvenile play-fighting can resemble dancing or boxing.

(one of the larger species) that repeatedly flew up to a stream of water coming from a hose, landed and "rode" the stream, then flew back up to ride it again.

Songbirds, woodpeckers, parrots, gulls, pelicans, hawks, crows, toucans, and penguins all belong to bird families that engage in locomotor, object, and social play. Among all of these playful birds, two of the most playful, and also best studied, are ravens and keas.

Ravens are large relatives of crows. These intelligent birds can manipulate things with their beaks and claws, and they have a vocabulary of calls for communicating with each other. Researchers like Bernd Heinrich, who has studied ravens in Vermont for many years, have seen them perform active group play, such as sliding on their backs down slippery banks of snow.

As you might expect for creatures with wings, ravens also play in the air. Young ravens sometimes band together for aerial acrobatics, diving, circling, and swooping in large numbers. Pairs of ravens engage in play chases and fights, in the air, on the ground, and even in trees. Ravens use objects in their social play, too. The birds hang upside down from branches and pass sticks back and forth. They hide objects such as pebbles or pieces of string and then try to find each others' objects, and they play "keep away" games in which one raven holds an object and another tries to snatch it.

Keas are parrots that are native to New Zealand. They are famous—or notorious—for their love of object play. Keas play with windshield wipers ripped from cars, covers taken from garbage cans, and other

For a young kea, even a traffic cone can be a toy. These New Zealand parrots are known for their playfulness, curiosity, and love of nibbling on things.

human-made objects not necessarily meant to be bird toys. With strong, sharp beaks and feet that can grasp and twist, keas can fashion playthings out of just about anything, including pieces ripped from outdoor furniture. Researchers once watched a kea repeatedly roll up a doormat and shove it down a staircase, apparently just for fun.

Social play is important in kea life. One of the most thorough investigations of social play in any wild animal was a 1999 study of keas by Judy Diamond and Alan B. Bond. Diamond and Bond documented the energetic, complex games that as many as six keas play together. They showed that the two main types of kea social play are tussle play, a rough-and-tumble wrestling activity, and toss play, in which the birds toss objects to each other. Because of research like this, scientists now know that keas, ravens, and other well-studied birds spend as much time playing, and play with as much complexity, as the most playful mammals.

Reptiles

Male Komodo dragons, the world's largest lizards, can reach lengths of almost 10 feet (3 meters). Although the dragons' normal food is the carcasses of dead animals, which they tear apart with their powerful jaws, they do attack live prey—including, on very rare occasions, humans. If the prey survives the attack and escapes, it is likely to die later of infection from the strong strains of bacteria in the dragon's mouth. In short, Komodo dragons are fearsome beasts.

But they are also playful.

Few reptiles have shown clear evidence of play, but there are many

Don't try this at home: Komodo dragons, dangerous reptiles that are not usually regarded as playful beasts, have shown some tendency to play when held in zoos.

reports of Komodo dragons playing with objects, both in captivity and in the wild. Kraken, a female dragon at the National Zoo, liked to push a bucket across the floor and play with objects including Frisbees, shoes, and boxes. She removed notebooks from her keeper's pockets and carried them around in her mouth without damaging them. (Fortunately for the keeper, deadly bacteria are absent from the mouths of zoo dragons, which are treated with antibiotics.) Kraken also entered into social object play, such as tug-of-war with her keepers.

THE CASE OF THE PLAYFUL ANTS

When a nineteenth-century Swiss scientist named Pierre Huber wrote that he had seen ants play-fighting, standing on their rear legs to wrestle without hurting each other, most people thought he had made a mistake. The ants he saw weren't playing—they were really fighting.

Around the same time, a different kind of ant play was reported by Henry Walter Bates, who spent years observing insects and wildlife along the Amazon River. Bates claimed that army ants, while marching through the forest in a long column, would sometimes stop in a sunny spot to rest, walk aimlessly around, and groom themselves or each other. He described their behavior as "idle amusement," "relaxation," and "merely play."

In 1905 a student of ant behavior wrote that on days of good weather, well-fed ants entertained themselves with "sham fights," like bear cubs romping or lambs frolicking. Other observers reported play-fighting, too, but most biologists and psychologists felt that calling such behavior "play" was too big a leap. The twentieth century's greatest ant expert, E. O. Wilson, made his position clear in a 1990 book with a section titled "Ants Do Not Play." The play-fights Huber and others had seen, Wilson said, were not play—they were fights.

We may never know whether an ant wrestling match without injuries is a serious fight or a play one. Researchers would have to design a study or experiment to show the difference. Modern ant researchers do know, however, that although ants are busy insects, they do not work all the time. They sometimes wander around with no known purpose, or hang out together in front of their nest. Maybe this behavior is relaxation or recreation. Maybe not. Science cannot yet tell us whether an ant feels pleasure.

A few observers have reported ants acting in ways that look like play-fights, or relaxation. Science is a long way from knowing, however, whether insects have behavior that can truly be called play.

Alligators, snakes, chameleons, fence lizards, and iguanas have all displayed behavior that might be play—but might also be exploratory curiosity or feeding, fighting, or mating behavior. Aside from Komodo dragons, the only reptiles that are known for certain to play are pond turtles. Some species of these freshwater turtles swim together in mock or play courtship, even when they are not mating.

Fish

Fish have caught the attention of animal behavior researchers in recent years. New studies have collected a growing body of knowledge about the complex social lives, cooperation, **cognition** or mental activity by fish, and even tool use by some species. There is also evidence that at least a few types of fish play.

Many species of fish jump out of the water. Sometimes they even jump over floating objects such as leaves and logs, as though playing leapfrog. The usual explanations for a fish's jump are that it is trying to escape from a predator or trying to catch a flying insect. The behavior has been recorded, however, at times when neither of those explanations could be correct.

In the 1890s a zoologist watched a number of garfishes take turns jumping over the back of a hawksbill turtle that was asleep on the surface of the water. This acrobatic display took place in an aquarium, with no predators present. The zoologist reported that the turtle appeared surprised and irritated when one of the garfish landed on its back and woke it up.

ANIMAL GAMES

An ichthyologist—a scientist who studies fish—holds an African elephant fish, one of the few fish known to play with objects, but only in captivity. No one knows yet whether wild fish play.

Jumping has been reported for garfish and other species right up to the present day. Some scientists think that fish jump to knock parasites off their skin. Others suggest that they jump to communicate. The sounds they make when they land on the surface travel through the water and may help members of a school keep track of each other's locations. It's possible, though, that fish sometimes jump for fun, in a form of finny locomotor play.

Object play has been recorded in the mormyrids, a group of African fish. One specimen, a captive elephant fish, liked to balance items such as twigs and even live snails on its snout. It also liked to play with balls, and after its keeper attached a small ball to the bottom of its tank with a string, the fish would frequently bat the ball back and forth. Similar behavior has been seen in other captive mormyrids and in cichlids, which are also African freshwater fish. It is not known if these fish play with objects in the wild.

Sharks may seem even less likely to play than Komodo dragons, but zoologist and shark expert Erich Ritter filmed a group of great white sharks playing with a bag of fish dangled over the side of a boat. The sharks shook and tugged at the bag even after they realized that they were not going to be able to eat it. They turned on their backs, raised their heads out of the water, and looked directly at the people on the boat, behavior that would not normally occur if the shark was fighting or attacking prey. Individual sharks came back to the boat repeatedly to play with the bag.

Scientific exploration of fish play is still in its early stages. To discover more about play in wild fishes of many species is going to be a challenging task. One important step will be finding foolproof ways to tell the difference between locomotor play and ordinary swimming, and between social play and being part of a school.

Animals Without Backbones

Nearly all studies of play have focused on vertebrates, or animals that have backbones. What about the invertebrates—the animals without backbones? This large group includes such creatures as insects, spiders, crabs and lobsters, worms, jellyfish, sponges, and starfish. The most fascinating discoveries so far concern octopuses.

Together with squids, cuttlefish, and nautiluses, octopuses make up a subgroup of invertebrates called **cephalopods**. The cephalopods have large brains for their size, and octopuses have become stars of animal behavior research because they seem to be the smartest invertebrates. They are active predators. They use their flexible arms to open shells and build shelters. Do these soft-bodied, eight-armed ocean dwellers also play?

They do, say a growing number of researchers who have studied captive octopuses. In one study, eight octopuses, each in its own tank, received a variety of small floating bottles. Most of the octopuses grasped the bottles and handled them. The handling was exploratory at first, then playful. The octopuses also moved the bottles around by squirting jets of water from their bodies.

Two of the animals went further. They squirted their bottles into the streams of water flowing into their tanks from fill hoses. The streams pushed the bottles back to the octopuses, which then squirted the

HOW ANIMALS PLAY

Captive octopuses, such as this one in a Spanish aquarium, often handle and even play with objects provided by humans. Whether these curious, highly intelligent animals play in the wild remains an open question.

bottles back to the hoses. The octopuses played this game repeatedly. In another study, an octopus kept grabbing a floating object, pulling it to the bottom of its tank, and releasing it, apparently to see it pop back up to the surface. Still another study showed that octopuses preferred red-and-white Lego blocks to bottles. The animals carried the blocks around with them inside their tanks.

Captive octopuses, living in security with a steady supply of food, play with objects that are provided by humans, but these may be special cases. Although scuba divers have reported anecdotes about playful behavior in wild octopuses, more research is needed before scientists know how much, if any, of a wild cephalopod's life is spent playing.

Octopus play is the frontier of an unexplored territory. Very little research has been done on play in other invertebrates. Is it even possible that a starfish or a worm could do something that we would recognize as play? Future researchers may discover where nature has drawn the limits of fun.

4. THE MEANING OF PLAY

Play comes with a cost. For one thing, playing takes energy, which may be of great value in a world of uncertain food supplies. Play can also be dangerous. An animal may hurt itself while playing, but a bigger danger is that it may catch a predator's eye, or get so involved in play that it doesn't see the predator sneaking up on it.

So why do animals play? What do they gain from it?

Scientists who have studied play in humans and other animals have put forward many theories about its purpose. No single theory is accepted by everyone, however. Different categories of play may have different purposes. Play may exist for a reason that no one has yet pinned down, or for more than one reason.

Early Ideas about Play

For most of the twentieth century there were three main theories about why play exists. Current theories about play draw on elements

Why are these red foxes playing? Scientists have come up with many different ideas about what animals—including humans—get out of playing.

from some of these early ideas, although the focus now is on the physical and mental effects of play.

The Long Arm of the Past

One early theory grew out of a study of play in humans. This theory is sometimes called recapitulation, which means giving a short version of what has gone before. It said that play was a leftover from the primitive past of the human species, a leftover of behavior that was once necessary. According to this theory, the children of today like to build sand castles and dig holes in sandboxes, and to play games that involve throwing balls and hitting them with sticks, because prehistoric humans built earthen dwellings, threw stones to kill game, and fought with clubs.

No one takes recapitulation seriously today. Scientists do think that the past is important to understanding play—but they mean the evolutionary past, the genetic links between today's living things and their long lines of ancestors, not the customs or habits of those ancestors.

High Spirits

A second early theory said that animals play when they have surplus, or extra, energy after their survival needs have been met. In other words, play uses up the physical and mental energy that hasn't been used for eating, hunting, foraging, escaping, or other necessary activities. In this view, play is an expression of energy, liveliness, and vigor—what some might call "high spirits."

THE MEANING OF PLAY

Instinct and Practice

The third major early theory said that animals are born with a play instinct—a drive that is genetically programmed. That instinct leads them to practice behaviors that they need in life. A mountain goat that leaps from one rock to another and back again, for example, is practicing for the leap that may one day save it from a mountain lion. Play-fighting ravens are sharpening the skills they will depend on to compete for mates or for food.

Excess energy, sometimes called high spirits, may be one reason animals play. Young mountain goats sometimes appear to leap and frolic from sheer excitement.

Benefits of Play

Energy and instinct continue to play a part in new views of play. The idea that play springs from energy is related to the notion of the relaxed field—a state in which an animal (or person) is well-fed, rested, and free from stress. Under such conditions, some kind of movement or activity may be natural to many, if not all, creatures. Should "play" be defined broadly enough to include any motions that are made just to release energy?

Instinct is central to much modern thinking about play. Animals that play seem to have an instinctive drive to do so, although they may have to learn specific play behaviors from their parents or siblings, or by watching other animals. Young children, for example, instinctively like to run around, but they must learn the rules of the game in order to play tag and hide-and-seek.

Why did a play instinct develop in the first place? What value does it have for animals? To answer those questions, researchers ask what benefits play brings to an animal.

Long-Term Benefits

Many species of animals play only when they are young. Or, if play continues throughout an animal's life, as it does with wolves, horses, and others, the adults play far less than they did when they were young. This seems to support the theory that play is a kind of education or training, in which young animals learn and practice the skills they will need later in life. Play, in other words, has a long-term benefit.

A RICHER LIFE

Pigface, the Nile soft-shelled turtle that lived in Washington's National Zoo for half a century, was one of the most playful reptiles yet known. He seemed to delight in playing with objects in his tank and with streaming water from a hose. Part of Pigface's story is *why* his keepers started giving him playthings.

The turtle was hurting himself, clawing his neck and biting into his front legs. His self-injuries were causing serious damage, and his caretakers thought that having objects in his tank might distract him and stop the destructive behavior. Their plan worked. Pigface took an interest in the toys, and he stopped hurting himself so much. From that time on his keepers kept giving him different objects to keep him from getting bored.

Pigface's self-injury was a kind of repetitive behavior often seen in captive animals. This behavior, which can be self-destructive and is repeated again and again without the slightest variation, is called **stereotypy**. It causes birds to pull out their feathers, bears to pace endlessly back and forth in a set pattern, rats to groom themselves obsessively until they bite off their own fur, and calves confined to small pens to roll their tongues hundreds of times an hour. Stereotypy may look like play, but it is solitary and compulsive. In humans, stereotypy is a sign of a mental disorder. In captive animals it seems to be brought on by extreme boredom, confinement in small or sterile environments, and lack of stimulation.

To prevent stereotypy, zoos now try to provide more stimulating environments for their animals. This may mean creating larger or more realistic habitats, in contrast to the barren cages of bygone years. Often it also means special programs for environmental enrichment.

At the Oregon Zoo, keepers hide fish and dry meat in hollow logs and other nooks in the enclosure of the Siberian tigers. The animals can smell the food and have to hunt for it. Elephants receive balls filled with treats such as nuts. When the animals play with the balls, bits of food fall out of small holes drilled in the balls. One of the most important tasks in a modern zoo is learning or creating new ways to enrich the lives of animals who, even in the best zoos, are living a life that is far from natural. For nearly all mammals and some birds, play is a vital part of enrichment.

HOW ANIMALS PLAY

The problem with this theory is that there has not been much research on how early play, or the lack of it, affects an animal's welfare later in life. Researcher Gordon Burghardt points out that it would be hard and perhaps cruel to carry out large-scale experiments, such as giving one group of kittens opportunities to play while preventing a second group from playing, then later seeing which adult cats are better at killing prey.

Limited information does suggest that early play improves later performance. When some young weasels were raised without a chance to play-fight, they had no opportunity to practice neck-biting. Later, as adults, they were not very good at two weasel activities that require neck-biting: killing rats (usually done with a fatal bite to the neck) and mating (the male weasel grips the female by the back of her neck).

Another long-term benefit of play, many scientists think, is **socialization**. This is the process through which a young animal—including a human animal—gains the skills and knowledge it needs to be part of a community or society. Leopards, sloths, and polar bears, among others, need little socialization. These solitary animals spend most of their lives alone, except for mating and child-rearing. Antelopes that live in herds, chimpanzees that live in troops, parrots that live in flocks, and naked mole rats that live in colonies are among the many species that need socialization because they will spend their lives interacting with others of their kind.

At the Wolong Nature Reserve in China, giant pandas can clamber about on a large wooden play structure. More and more zoos recognize the need to provide captive animals with mental and physical stimulation.

There is little scientific data about whether play truly improves an animal's long-term social success, such as its chances of mating or becoming dominant in a group. Still, Marc Bekoff, an ethologist who has studied wolves, coyotes, and dogs, argues that play helps young animals in these species acquire social awareness. Through play, young wolves learn to interpret the eye movements of other wolves, and this helps them predict what the others will do. Play also gives the animals experience in dominant and subordinate roles. Over time, social play and play-fighting help determine who will occupy each rung of the social ladder.

Short-Term Benefits

One feature of play listed in the first chapter of this book is that play has no immediate purpose or goal. It gives the animal no survival advantage at that moment. A baby elephant trunk-wrestling with an older sister is not striving for something necessary to survival, such as food or escape from a predator. It's just playing.

Yet play may have short-term benefits that are important but invisible. Research suggests that physical activity at a crucial stage of a young animal's growth may be linked to the development of its body and brain. The first part of this research was to discover how play is spread out or concentrated in an animal's lifespan. This has been documented for some species. Olive baboons, for example, reach their peak play rate—nearly five minutes of social play every hour—at about sixteen months of age. At forty months old the baboons are engaged in social play for less than two minutes an hour, and after that the rate of play drops off sharply.

Biologist John A. Byers looked at play rates in the lives of three species: house mice, Norway rats, and domestic cats. He saw that their times of maximum play exactly matched an important period in their biological development—the point in the animal's growth when certain nerve fibers form in the brain, and muscles throughout the body become either fast-twitch muscles (good for quick reactions) or slow-twitch muscles (good for strength). The more experiences an animal has during this period, and the more physical movements it performs, the better its nerve fibers and muscles will develop. And what gives a young animal more experiences and movements than playing?

The play switch is turned on, as Byers puts it, right at the time when experience can build fibers in the brain and movement can set the balance of fast- and slow-twitch muscles. When that sensitive period of development ends, the play switch is turned off, and animals stop playing, or they spend much less time playing. Byers thinks that play exists, at least in part, to help shape brains and bodies.

The Love of Play

In 1851 Edward P. Thompson explained why he had written a book about animal behavior, including play. He wanted to awaken a "kindly feeling" toward animals, so that his readers would give the animal world "the admiration and protection" it deserved. Since Thompson's

HOW ANIMALS PLAY

Romping, rolling, and sliding in the snow may help this young polar bear's muscles and nerves develop, but it looks like fun as well. People also play for both pleasure and benefits such as exercise. Playfulness is something that humans share with many of the world's other creatures.

time the study of play has given us insights into the animal world to which we belong. It has also shown that play is an important part of welfare for many animal species, perhaps more than we know.

Human toddlers push toy trucks. Teenagers shoot hoops and rack up high video game scores. Grown-ups swat tennis balls—or relax and watch someone else do it. It is clear that people like to play, when they have the time and resources for it. This play instinct is something that we have in common with other members of the animal world. Maybe those other animals also experience something of the pleasure and joy that we take in play.

GLOSSARY

anecdotal	Based on anecdote, or someone's account of what happened; it is not strong scientific evidence because it cannot be tested.
anthropomorphism	Thinking that animals' features, characteristics, and reactions are like those of humans; interpreting animals in human terms.
archosaurs	Four-limbed, egg-laying animals that evolved around 250 million years ago.
biology	Scientific study of living things.
cephalopod	Member of a group of sea animals that includes octopuses and squids.
cetacean	Member of the group of sea mammals that includes whales, dolphins, and porpoises.
cognition	Mental activity; act of thinking.
crocodilians	Group of large reptiles that includes living and extinct crocodiles, alligators, gharials, and caimans.
dominant	Having higher rank, status, and power than others.
ethogram	Chart or list of the various behaviors of an animal that is being studied; shows how many times the animal performs each type of behavior and for how long.
ethology	Scientific study of animal behavior.
evolution	Process by which new species develop over time because of changes, or mutations, in existing species.
forage	Look for food.

GLOSSARY

habitat	Environment in which an animal naturally lives.
instinct	Pattern of behavior that is genetically programmed, and that all members of a species are born with.
locomotor play	Play that consists of locomotion, or moving the body, as in running, jumping, pouncing, or bouncing.
locomotor-rotational play	Play that combines twisting or turning movements with locomotion.
object play	Activity in which an animal (or person) plays with one or more objects.
parallel play	When two animals play near each other and appear to be aware of each other, but do not interact directly.
predator	Animal that preys on, or hunts and eats, other animals.
predatory play	Play that mimics, rehearses, or uses movements from hunting.
prey	Animal that is hunted or killed by a predator.
primates	Group of animals that includes monkeys, apes, and humans.
psychology	Study of the mind; comparative psychologists study the differences and similarities between human and animal minds.
self-handicapping	Playing with limited energy, strength, or skill in order to remain "equal" with a play partner that is smaller or weaker.
social play	Play in which an animal (or person) plays with one or more other animals (or people).
socialization	Process by which a young animal learns how to interact and get along with others of its kind, to be part of a community or society.
sociobiologist	A scientist who studies how animals that live in social groups, such as ants and zebras, interact with each other.
species	Group of plants or animals that are enough like one another to have offspring that are fertile (able to produce offspring of their own).
stereotypy	Repetitive, sometimes self-damaging behavior seen in some captive animals, caused by sterile or small environments and lack of stimulation.
subordinate	Having lower rank, status, and power than others.
ungulate	Mammal with hooves.
zoology	Branch of biology that studies animals, including insects.

FIND OUT MORE

Books

Bekoff, Marc. *Animals at Play: Rules of the Game.* Philadelphia: Temple University Press, 2008.

Bekoff, Marc, ed. *The Smile of a Dolphin: Remarkable Accounts of Animal Emotions.* New York: Discovery Books, 2000.

Boysen, Sally. *The Smartest Animals on the Planet.* Buffalo, NY: Firefly, 2009.

Tatham, Betty. *How Animals Play.* New York: Scholastic, 2004.

Websites

Environmental Enrichment
The Oregon Zoo, a leader in studying how enrichment—including play—can improve the lives of captive animals, maintains this page with links to examples of enrichment programs for a variety of creatures, from polar bears to birds.
www.oregonzoo.org/Cards/Enrichment/conserv_enrichment.htm

Jungle Gyms: The Evolution of Animal Play
The National Zoo in Washington, D.C., posts this article about animal play—which animals play, what purposes play serves, and how play might have evolved.
http://nationalzoo.si.edu/Publications/ZooGoer/1996/1/junglegyms.cfm

Mind & Brain/Animal Intelligence
Discover's Animal Intelligence page has links to dozens of articles about recent discoveries, written for ordinary people, not scientific experts.
http://discovermagazine.com/topics/mind-brain/animal-intelligence

Zoo Animal Enrichment
The National Zoo's website has information about how zookeepers can provide animals with opportunities to play and perform other actions that keep them active, alert, and interested in life.
http://nationalzoo.si.edu/SCBI/AnimalEnrichment/default.cfm

BIBLIOGRAPHY

The author found these books and articles especially helpful.

Balcombe, Jonathan. *Pleasurable Kingdom: Animals and the Nature of Feeling Good.* New York: Macmillan, 2006.

Bekoff, Marc. *Animal Passions and Beastly Virtues.* Philadelphia: Temple University Press, 2006.

———. *The Emotional Lives of Animals.* Novato, CA: New World Library, 2007.

Bekoff, Marc, and Jessica Pierce. *Wild Justice: The Moral Lives of Animals.* Chicago: University of Chicago Press, 2009.

Bekoff, Marc, and John A. Byers, eds. *Animal Play: Evolutionary, Comparative, and Ecological Perspectives.* Cambridge, UK: Cambridge University Press, 1998.

Burghardt, Gordon M. *The Genesis of Animal Play: Testing the Limits.* Cambridge, MA: MIT Press, 2005.

———, Brian Ward, and Roger Rosscoe. "Problem of reptile play: Environmental enrichment and play behavior in a captive Nile soft-shelled turtle, *Trionyx triunguis*." Zoo Biology 15, no. 3 (1996): 223–238.

Burkhardt, Richard W., Jr. *Patterns of Behavior: Konrad Lorenz, Niko Tinbergen, and the Founding of Ethology.* Chicago: University of Chicago Press, 2005.

Diamond, Judy, and Alan B. Bond. *Kea: Bird of Paradox.* Berkeley: University of California Press, 1999.

Flack, Jessica C., Lisa A. Jeannotte, and Frans B. M. de Waal. "Play Signalling and the Perception of Social Rules by Juvenile Chimpanzees (*Pan troglodytes*)." *Journal of Comparative Psychology* 11, no. 2 (2004): 149–159, www.emory.edu/LIVING_LINKS/pdf_attachments/playsignals_JCP04.pdf

Griffin, Donald R. *Animal Minds: From Cognition to Consciousness.* 2nd ed. Chicago: University of Chicago Press, 2001.

Hatkoff, Amy. *The Inner World of Farm Animals.* New York: Stewart, Tabori & Chang, 2009.

HOW ANIMALS PLAY

Heinrich, Bernd. *Ravens in Winter.* New York: Vintage, 1989.

Lovgran, Stefan. "Dinosaurs Were Doting Parents, Fossil Find Suggests." *National Geographic* News. September 8, 2004, http://news.nationalgeographic.com/news/2004/09/0908_040908_dinoparents.html

McCarthy, Susan. *Becoming a Tiger: How Baby Animals Learn to Live in the Wild.* New York: HarperCollins, 2004.

Sarti Oliviera, Ana Flora, and others. "Play behaviour in nonhuman animals and the animal welfare issue." *Journal of Ethology* 28 (2010): 1–5.

Uhlenbroek, Charlotte, ed. *Animal Life.* New York: Dorling Kindersley, 2008.

Wynne, Clive. *Do Animals Think?* Princeton, NJ: Princeton University Press, 2004.

———. *Animal Cognition: The Mental Lives of Animals.* New York: Palgrave, 2001.

INDEX

Page numbers in **boldface** are photographs.

age, 12–16, **15**, 66, 68, 70–71
anecdotal, 17, 45, 47, 61
anthropomorphism, 8–9
archosaurs, 24

behavior questions, key, 8
benefits of play, 66–72
birds, 6, **7**, 24–25, **25**, 30, 41, 48–52, **51**, 67
bonding, 6, **13**, **21**
bullying, 33, 37–38, **37**

carnivores, 45
cephalopod, 59, 61
cetacean, 45–46
classification system, **26**, 27
cognition, 56
crocodilians, 24
cross-species play, 34, **35**

Darwin, Charles, 14, 17
dinosaurs, **10**, 11, 23–24, **23**, **25**
dominant, 38, 70
dopamine, 20

ethogram/ethology, 6, 9, 43, 44
evolution, 9, 14, 24–25, 64

fast-/slow-twitch muscles, 71
fish, 27, 56–59, **57**
fossils, 23–24, **23**

group play, 50

habitat, 16, 67
high spirits theory, 64–66, **65**

instinct, 6, 8, **13**, 65, 66, 73
invertebrates, 59–61, **60**

Kraken, 53

locomotor play, 27–28, **29**, 38, 40, 43, 46, 50, 58–59
locomotor-rotational play, 28
Lorenz, Konrad, 6, **7**, 8

mammals, 28, 30, 41, 45–48, 52, 67
marsupials, 46, 48, **49**
metacommunication, 39

object play, 27, 28, 30–33, **32**, 38, 43, 46, 48, 50, 53, **57**, 58, **60**, 61, 67

parallel play, 38, 41
Pigface, **42**, 43–45, 67
pinnipeds, 46

79

play, 12–16
  early theories, 64–65
  five features of, 18–23
  origin of, 23–25
play-fighting, 20, **21**, 36, 48, **49**, 54, **55**, 65, 68, 70
predator, **4**, 6, **10**, 18, 30, 34, **35**, 56, 59, 63, 70
predatory play, 30, **32**, 33
prey, 14, **19**, 30, 32, **33**, 34, 52, 58, 68
primates, 30, **37**, 38, 39, 45
proboscids, 45

recapitulation theory, 64
relaxed field, 22–23, 54, **55**, 66
reptile, **42**, 52–53, **53**, 56, 67

rodents, 46
role reversal, 36

self-handicapping, 36, 48
socialization, 68
social play, 33–38, **35**, 40, 43–44, 46, 48, 50, 52–53, 59, 70–71
species, 6, 8, 14, **15**, **23**, 30, 34, 36, 38, 41, 45, 46, 48, 56, 58, 64, 66, 68, 70–72
stereotypy, 67
subordinate, 38, 70

toy, 22, 8, **30**, **32**, 40–41, 44, **47**, **51**, 52, 67

ungulate, 28, 46

REBECCA STEFOFF has written many books about animals for young readers of all ages. Her book *Ant* (1998) was made into a chapter in a popular reading textbook for second graders. More recently Stefoff wrote ten books, including *Horses*, *Penguins*, *Chimpanzees*, and *Tigers*, for young adults in the AnimalWays series. For the same publisher's Family Trees series, she explored twelve groups of living things, from *The Fungus Kingdom* to *The Primate Order*. Stefoff lives in Portland, Oregon, where she enjoys bird watching, kayaking, and visiting the zoo. You can learn more about her and her books for young people at www.rebeccastefoff.com.